The Delicacy of Balance

A cosmic journey through myth, science, and poetry

Christopher Cotton

Published and Manufactured by Softwood Books

EU Responsible person: Maddy Glenn
Office 2, Wharfside House, Prentice Road, Stowmarket, Suffolk, IP14 1RD
www.softwoodbooks.com, hello@softwoodbooks.com

EU Rep:
Authorised Rep Compliance Ltd., Ground Floor, 71 Lower Baggot Street, Dublin, D02 P593, Ireland
www.arccompliance.com, info@arccompliance.com

A CIP catalogue record for this book is available from the British Library

Paperback ISBN: 978-1-9192021-1-2

This book is dedicated to:

Mandy – my wife, my north star.

Suffolk Writers Group – my constellation of inspiration.

Open minded, curious folks everywhere – you are our way-finders.

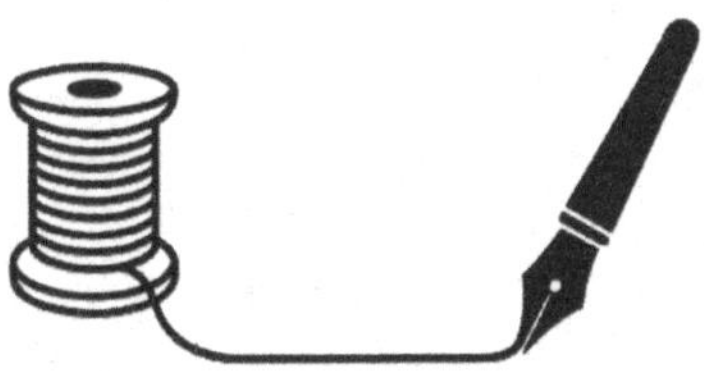

"If you look at Earth from space you see a dot, that's here. That's home. That's us. It underscores the responsibility to deal more kindly and compassionately with one another and to preserve and cherish that pale blue dot, the only home we've ever known."

Carl Sagan

"The fact that we live at the bottom of a deep gravity well, on the surface of a gas covered planet going around a nuclear fireball 90 million miles away and think this to be normal is obviously some indication of how skewed our perspective tends to be."

Douglas Adams

"Wheels within wheels in a spiral array
A pattern so grand and complex
Time after time we lose sight of the way
Our causes can't see their effects."

Neil Peart

Contents

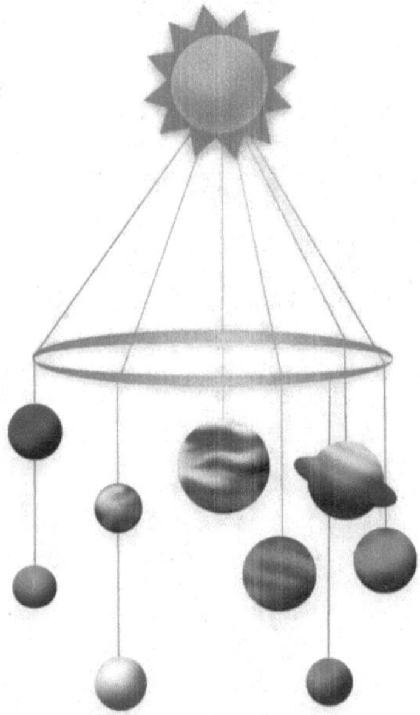

Foreword

A long time ago, when I was training in Family Therapy, one of my tutors used a striking metaphor: a child's mobile. His version had a dozen gaudy wooden parrots, each suspended by string. The system held together just as long as every bird was in balance; shift the weight of just one, and the whole mobile tilted, forcing everything else to adjust until a new equilibrium was found.

That image has stayed with me. It is a simple way of showing that no system, family, organisation, or society, can be changed in one part without consequences elsewhere. Balance is never static; it is a continual negotiation.

The same truth applies far beyond the walls of a therapy room. Astronomers now suspect there may be a "ninth planet" (*Planet X*) tugging on the orbits of our known solar companions. Our own system is a dance of gravitational pulls and counter-pulls, each planet's weight felt by every other. And this solar system, in turn, is just one participant in the greater rhythm of the Milky Way, circling a supermassive black hole (*SGR A*) at its centre.

It may sound abstract, even grandiose, but it is the same principal writ large. Change one element, however distant, and balance shifts across the whole.

As human beings, the question of balance is never far from us. We share this small planet with countless other lifeforms, yet we also look outward and ask: *Is there anybody out there?* Statistically, the odds suggest there must be. Practically, it is unlikely that we will ever meet. So, we turn inward, and ask how we came to be here, how we understand our place in a universe both fragile and vast.

Every culture has origin stories, myths, faiths, cosmologies, that sought to explain what science has only more recently begun to map. In all of them, a tension runs between what we *believe* and what we *know*, between the stories that give meaning and the facts that give measure.

This book is an attempt to walk that tension line. We will trace the paths from myth to mathematics, from constellations as gods to constellations as charts, from stories of creation to the equations of physics. At each stage, I will offer a poetic reflection: not as an expert in astronomy, religion, or philosophy, but as a fellow traveller, curious about our world, our solar system, our galaxy, and our universal neighbourhood.

If what follows leaves you with a little more wonder, a little more appetite to look up at the night sky, or to look again at the myths that shaped us, then I will have achieved my purpose.

Christopher Cotton
Suffolk, 2025

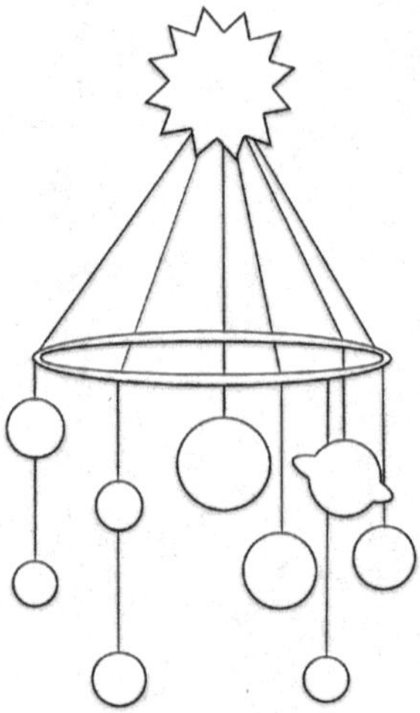

Prologue:

"In the beginning he established the balance and set the stars in their courses."
Enuma Elish

"The fault, dear Brutus, is not in our stars, but in ourselves, that we are underlings."
Shakespeare, Julius Caesar

"Equilibrium is the profoundest tendency of all matter."
Goethe

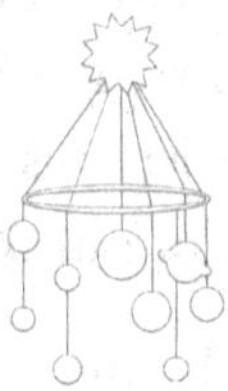

We have always lifted our eyes to the heavens, drawn by the mystery and majesty of the night. Creation stories fill the sky with chariots and messengers, gods and celestial beings, all struggling to wrest order from chaos, to bind the wild into pattern and predictability.

Yet in every story there is always a hand beyond our own: some god who set the spheres in motion.

This gift of imagination has been humankind's key, to conjure stories that explain our existence, our purpose, our rights. We are the great storytellers, binding ourselves together (and sometimes pulling apart) through the myths we have woven since first we wandered the earth in search of home.

From story came faith, from faith came scripture, and from scripture whole systems of order. "As above, so below" became our mantra, as we mirrored divine architecture in the heavens: halls, palaces, cathedrals, kingdoms… projections of our own striving.

We traced the stars into hunters, goats, fish, dragons, families, crabs. Astrology turned the night sky into a chart of fortune: to foretell battles, harvests, journeys, and love.

Yet still we thirsted for proof. Gradually knowledge displaced belief, though belief still binds our cultures together. With mathematics and observation, we dared to go further, beyond the myths of our ancestors. Spectroscopy let us read the colours of distant suns and know their composition.

Newton's *Principia* gave us the tools to measure not only the fall of an apple but the great motions of the spheres.

Throw an apple upward and it returns to earth. We live inside gravity's pull, even before we knew how to name it.

That unseen tethering force draws all things together, the greater the mass, the stronger the pull. Here it means an acceleration of 9.807 metres per second squared. Archers, gunners, and rocket scientists alike learn the same principle: what rises, falls, and always in a calculable arc.

But why then does the moon not crash into the earth? Why do the planets not tumble into the sun?

The answer is orbits. The moon is always falling toward us… and always missing. Its sideways motion carries it just far enough that the earth slips continually beneath it. So, with the planets and the sun: everything is falling, and everything is spared by motion.

This is the delicate balance we inherit. Stars, planets, moons, asteroids, comets, even dust: all exert their gravity, pulling and tugging, shifting their neighbours. Some fall inward to fiery ends; others find a rhythm, a stable arc, a fragile equilibrium.

Our solar system, like a mobile hung on string, is stable only while no thread is cut, no weight added. In time, our sun will swell, collapse, and remake its children. But that end lies far beyond our span. For now, the tension between falling and drifting, between chaos and order, holds.

*****

Balance

"Equilibrium is the profoundest tendency of all matter."
Goethe

We looked in wonder at the skies,
the heavens sweeping slowly by,
a steady drift of light through dark,
the turning wheel, the endless arc.

We placed ourselves as centre, true…
what else was there for us to do?
The ground was still, the seas at rest,
and so our vision seemed the best.

But time revealed the deeper math;
the orbits traced a different path.
The centre shifted… never ours,
we circle round far greater powers.

Now we live with delicacy:
the fragile art of falling,
the constant tug of gravity,
forever, always calling.

Like dominoes aligned in time,
one tremor starts the fall.
Which toppling shapes the story's arc,
and will it end us all?

*****

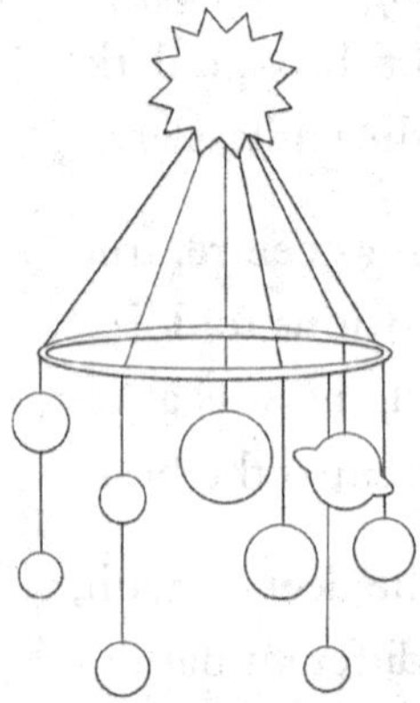

We, and all other lifeforms on this planet are part of a dynamic universal system, everything connects

Sol

"Helios, lord of the sun, watch over us and see that we walk in truth."
Homeric Hymn to Helios

"The sun… can still ripen a bunch of grapes as if it had nothing else in the universe to do."
Galileo

"Even after all this time, the sun never says to the Earth, 'You owe me.' Look what happens with a love like that, it lights the whole sky."
Hāfez

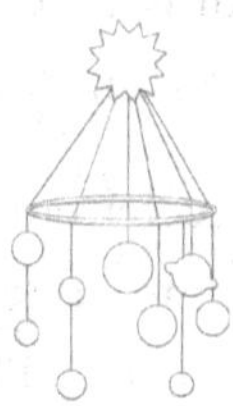

The Sun has been a central figure in mythologies across cultures, often revered as a deity embodying creation, life, and authority.

For the Egyptians he was Ra, sustaining life and maintaining cosmic order. Depicted as a man with the head of a falcon crowned by a solar disc, he voyaged each day across the sky in a barque of light, and each night through the underworld.

To the Greeks he was Helios, the sun itself, steering a golden chariot across the heavens, while Apollo, another solar figure, came to represent enlightenment and artistic inspiration.

In ancient Mesoamerica, the Inca worshipped Inti, ancestor of their people, while the Aztecs told how Nanahuatzin sacrificed himself to become the fifth and current incarnation of the sun.

In Japan, Amaterasu was celebrated as ruler of heaven. When she once retreated into a cave, the world was cast into darkness until the gods lured her out again. In Chinese legend, the goddess Shiho was mother to ten suns. When her children threatened to scorch the earth, an archer was sent to shoot down nine, leaving the one we know today.

The Norse also imagined the sun as female, driving her chariot endlessly across the sky, pursued by a ravenous wolf determined to devour her… a symbol of the fragile balance between light and dark.

Across all these myths runs a common theme: balance. The tension between light and shadow, presence and absence, order and chaos. Both are necessary; both sustain the world.

Beyond myth, our sun is the heaviest body in the solar system. At 1.989×10^{30} kg - 333,000 times the mass of Earth - it consists mostly of hydrogen and helium, with traces of oxygen, carbon, neon, and iron, all in a searing plasma state. Its immense gravity binds the core into perpetual nuclear fusion, releasing energy and creating helium in the process.

Sol, fulcrum of our solar system, the central pin in our mobile of planets.
Sol, holding the heavens in orbital balance and order, radiating both deadly winds and life-giving warmth.
Sol, our cosmic heartbeat… our eternal pulse.

*****

Sol Invictus

"The sun… can still ripen a bunch of grapes as if it had nothing else in the universe to do."
Galileo

For centuries you watched me rise and fall,
And thought me slave to Earth's revolving throne.
You traced my arc in myth and polished stone,
A burning god obeying nature's call.
Your rites were written in the shadows' sprawl,
But then your heresies became your own,
You found I held my place while you were thrown
In orbits vast, adrift within it all.

I am no god, no keeper of your fate,
Just one more star among a billion more.
My fire will fade, your time will come of late,
And neither crown nor creed will hold the floor.
You measure me to understand your state,
Yet I, unmoved, outlast your mortal war.

*****

Entropy

"Stars die so that the rest of the universe can live."
Philip Pullman

Our sun, though not alive, still hungers,
its banquet drawn from hydrogen,
a slow unspooling certainty.
What seems eternal brilliance
is but a spark in cosmic night,
youth beneath the gaze of time.

One day the balance shifts:
helium ascends,
and the furnace swells beyond its hold.
Gold turns crimson,
light becomes a consuming fire,
planets once cherished
fall into its breath.
The orbits end.

Yet, do not despair, dear traveller.
Billions of years lie between
this moment and that horizon.
By then our voices will be echoes,
our cities ruins beneath forgotten sands,
our monuments weathered into silence,

our wars, our wealth, our striving
no more than motes of dust
adrift in the scales of time.

But here, now,
balanced in the narrow grace of distance,
life is kindled.
What privilege,
to know existence at all,
to stand in the light of a star
and name the gift enough.

*****

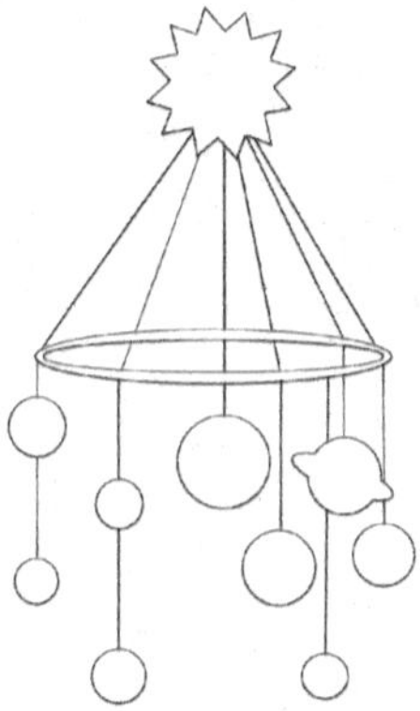

The Sun holds all in balance,
our first measure, our last horizon.

Mercury

‘Hermes, messenger of the gods, son of Maia, guide of souls to the underworld.”
Homeric Hymn to Hermes

‘Mercury’s orbit is the most capricious of all: tugged by Jupiter, drawn by the Sun, its path is a delicate balancing act.”
James Jeans

‘Fleet-footed Hermes, bringer of dreams, deceiver of men.”
Homer, Iliad

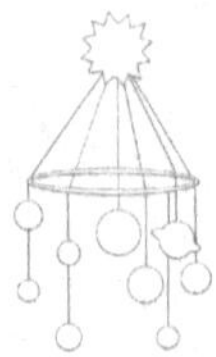

Mercury has always stood for speed. In myth he was the messenger - winged sandals flashing, darting between gods and mortals. In the heavens he is the planet that runs closest to the Sun, swiftest of all, tracing its orbit in just eighty-eight days.

At 57.9 million kilometres from the Sun's corona, Mercury seems distant by human measure, bound as we are to horizons and eyesight. Yet it is the innermost planet, the first dancer in the solar fire. To the naked eye he is elusive, glimpsed only at dawn or dusk, slipping quickly from view. The ancients thought him two stars, Hermes at evening, Apollo at dawn… until they discovered he was one.

Between Roman and Greek pantheons, Mercury inherited domains of commerce, boundaries, communication, and trickery. Speed, wit, and caprice were his attributes. But swiftness hides severity. His day is a furnace, his night an abyss. With no air to hold warmth, heat falls away into unimaginable cold. No clouds, no weather, no shelter. Just a world stripped bare, its surface pummelled by stone and star, locked in a dance too close to the fire.

With a mass of 3.3×10^{23} kg - about 5.5% of Earth's - Mercury is held in orbit by the Sun, yet tugged by other giants, especially Jupiter. Its path is fragile: some astronomers have suggested that, over vast timescales, the planet could be flung into outer space or drawn into collision with Venus or Earth. For now, balance holds, and stability continues.

The Romans, with some irony, named him son of Jupiter and Maia, guide of souls to the underworld. Mercury is a rapid messenger, but never a home. A fleeting glimpse… and gone, always on the edge of ejection or return to his father.

Mercury, embodiment of speed.
Mercury, planet of extremes.
Mercury, held in the most fragile balance.

*****

Fleet

"Fleet-footed Hermes, bringer of dreams, deceiver of men."
Homer, Iliad

At dawn, a quick spark,
the god with winged feet departs
night swallows the flame.

*****

Desiccate

"Mercury's orbit is the most capricious of all: tugged by Jupiter, drawn by the Sun, its path is a delicate balancing act."

James Jeans

Beneath the sandals
lies no garden,
no place for messages to rest.

Only rock,
baked to silence by day,
frozen to stillness by night.

The messenger runs on
and leaves us nothing.

*****

Caprice

"Hermes, messenger of the gods, son of Maia, guide of souls to the underworld."
Homeric Hymn to Hermes

O fleeting, racing, trickster god,
wing-shod herald of the skies,
your silver tongue and cunning wiles
once ferried messages of might,
but now you wear a rocky guise,
a fragile world too small to grow,
forever seared by furnace flame,
then frozen in your shadowed wane.

Within your year of eighty days
you reel through heat and darkness' maze,
a path so fine, so frail, so fast…
one slip, and order cannot last.
The balance falters, orbits fall,
and chaos seethes throughout it all.

No prayer, no rite, no human plea
could sway the fate prepared for thee.
We trust that in the scale of time
your caprice will not break the whole,
nor end us in so swift, so strange,
so undeserving a fall.

*****

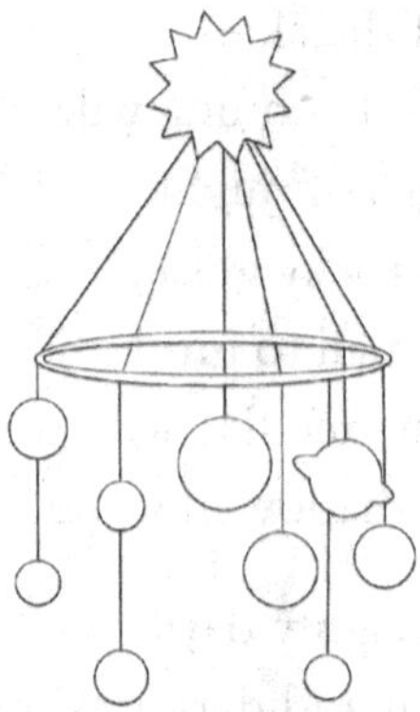

Swift messenger scorched and silent;
yet still a herald of the greater journey.

Venus

"Venus herself, with all her beauty, is also the cause of war and ruin."
Ovid

"Venus is a hell-world: its surface hot enough to melt lead, its clouds rain acid, its pressure like the ocean depths."
Carl Sagan

"She walks in beauty, like the night
Of cloudless climes and starry skies."
Byron

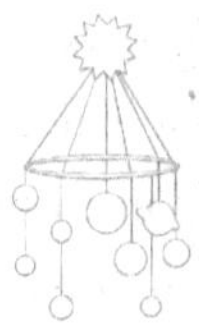

To the ancients, Venus was the brightest jewel of the heavens… morning star, evening star, the goddess of beauty herself. She rose radiant at dawn and dusk, a symbol of desire, fertility, and harmony. Calendars and festivals beat to her rhythm.

Steeped in mythology, Venus was named for the Roman goddess of love and beauty, a counterpart to the Greek Aphrodite. In Babylon she was linked to Ishtar; in China, the *Golden Star*, bringer of fortune.

Born, so myth tells us, from sea foam; Uranus's severed genitals cast upon the waves… she emerged fully grown and radiant, as Botticelli's painting immortalises.

But beauty deceives. The planet named for love is no paradise. Beneath her bright veil of cloud lies a world of suffocation: carbon dioxide thick as oceans, pressure ninety times our own, heat fierce enough to melt lead. Clouds of acid drift above a surface restless with volcanoes.

At 108 million kilometres from the Sun, Venus is almost Earth's twin in size and mass - 4.87 × $10^{\wedge 24}$ kg, about 81.5% of Earth. A Venusian year lasts 224.5 days. Geologically, she resembles us. Atmospherically, she is our antithesis.

From afar, she gleams. Up close, she destroys. Venus is the mirror of our longings… and the warning of what unchecked fire and runaway warming can bring.

Venus, the radiant, beautiful goddess.
Venus, the personification of Hell.
Venus, our salutary warning.

*****

Goddess

"A thing of beauty is a joy forever."
John Keats

She walks the dusk in silver,
veiled in softest light,
the goddess of desire,
who promises delight.

The shepherds name her radiant,
the poets call her flame.
She rises, pure and beautiful,
and whispers us her name.

A morning star of brilliance,
jewel-like, aflame…
but do not trust her beauty
as a sign that she is tame.

For gods can strike as well as bless,
and Venus grants no redress.

*****

Hellscape

"Through me you enter into the city of woe… Abandon all hope, ye who enter here."
Dante, Inferno

Peel back the clouded veil,
and beauty's face dissolves.

No garden waits,
no gentle shore.

Only pressure…
crushing bone and steel alike.
Only acid rain,
and oceans boiled away.

The goddess of love
is a furnace in disguise.
She waits to teach us
what happens
when the air itself betrays.

*****

Hothouse

We do not inherit the Earth from our ancestors; we borrow it from our children."

Native American proverb

An atmosphere of CO_2,
with N_2 to swell the air,
SO_2 drifting, H_2O misting,
trace elements scattered there.

This heady atmospheric soup
not only traps the heat,
it brews H_2SO_4...
to make the hell complete.

Corrosive veils of acid cloud,
a burning, choking sky...
the greenhouse of all greenhouses,
where even hope would die.

And if, in some forgotten age,
life flickered in that flame,
it's long since lost, erased, consumed,
no echo to its name.

Our sister in so many ways,
so close, yet scarred, afar:
a mirror to our fevered Earth,
a warning of what we are.

*****

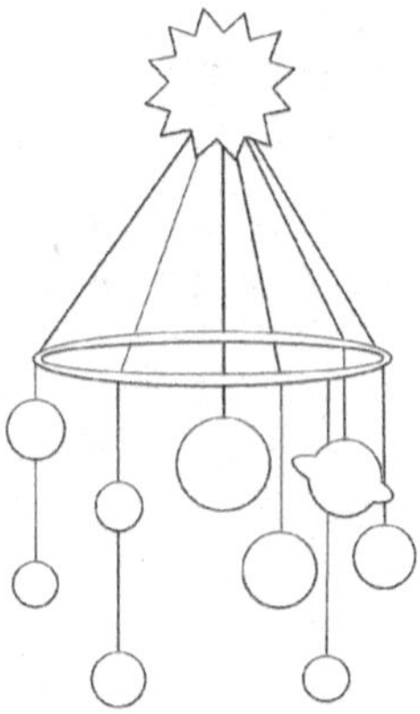

She burns with beauty and with warning,
a mirror of desire and of consequence.

Earth

"Gaia, mother of all, the firm foundation of the immortal gods and of man."
Homeric Hymn to Gaia

"The Earth is the only world known to harbor life. There is nowhere else, at least in the near future, to which our species could migrate."
Carl Sagan

"The Earth is what we all have in common."
Wendell Berry

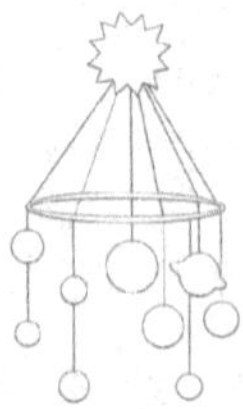

Earth: our cradle, our anchor, our only home. Unlike Mercury's silence or Venus's fire, this planet is tempered by balance. Rock and water, air and life… each held in fragile proportion.

Often imagined as a mother goddess, Earth appears in countless myths. Gaia, Terra, Pachamama, Asase Yaa, and Geb are but a few of her names. Creation stories tell of her arising on the back of a turtle, or from the body of Ymir, a slain giant. Whether as foundation or transformation, Earth has always embodied the generative.

For much of history we assumed she was the centre of all things. It felt true: the solid ground beneath our feet, the heavens turning above. But we learned otherwise. We are not the centre… only one world orbiting a modest star.

Curiously, astrologers rarely chart Earth in their predictions of the divine. Yet if we are not central, we remain exceptional. This is the one place we know that harbours life. A thin film of atmosphere guards us; oceans circulate warmth; tectonic plates shape continents; photosynthesis has painted the skies with breathable air.

Earth orbits the Sun at about 150 million kilometres, safely within the "Goldilocks zone" - not too hot, not too cold. With a mass of nearly six trillion, trillion kilograms, she is the largest rocky planet, and the first with enough gravity to hold a permanent satellite: the Moon.

Her molten core, magnetic shield, tides and circulations are the rhythms of our world. But balance is never guaranteed. Our abundance tempts us to excess; our ingenuity risks our survival. Earth is resilient, but she is not invulnerable. To live here is to remember the fragility of her equilibrium… and the impossibility of ever finding another quite the same.

Earth, our mother.
Earth, our home.
Earth, our priceless jewel.

*****

Bosom

"The Earth does not belong to us: we belong to the Earth."
Chief Seattle

O turning sphere,
clothed in ocean and cloud,
veined with rivers,
crowned in green.

The only hearth we know,
where fire warms,
where water sings,
where breath is shared.

Life suckles at your bosom,
multiplied in myriad forms,
evolving, ever changing
to fit the world we know.

O Earth,
we are your children,
and your wonder is our home.

*****

Gossamer

"The Earth is a very small stage in a vast cosmic arena."
Carl Sagan

Above us lies a membrane
thinner than thought.
Beneath it we build,
beyond it the void.

A delicate balance
of chemistry and flow:
systemic dancer,
cyclical marvel.

We assume its permanence
and squander the gift.
One tremor of balance,
one fever unchecked,
and the skin tears.

Oceans rise.
Forests fall silent.
What was cradle
becomes coffin.

*****

Sanctum

"There is nothing like staying at home for real comfort."
Jane Austen

I have walked your hills,
touched your stones,
breathed your morning frost.

I have carried your soil in my hands,
and your rain upon my face.

I know no other world,
nor want one.
I know no other dawns,
nor other suns.

If I must fall,
let me fall here…
in the place
I call home.

The place where I belong.

*****

Luna

"The moon is a friend for the lonesome to talk to."
Carl Sandburg

Companion moon, you circle Earth in space,
our tethered hearts entwined in silent turn.
You mirror back the longing of her face,
and pull her seas with tides that ebb and yearn.

They named you goddess, pale of skin, divine,
a keeper of the sacred mark of time.
You glimmer, carved in stone and whispered line,
eternal… though you slowly drift, sublime.

Yet myths endure, though knowledge will not sever:
each year you inch away, a measured flight.
Your borrowed glow still binds our gaze together,
and draws our wonder to the edge of night.

Though you may go, our dance has shaped endeavour,
and cast our eyes beyond to seek new light.

*****

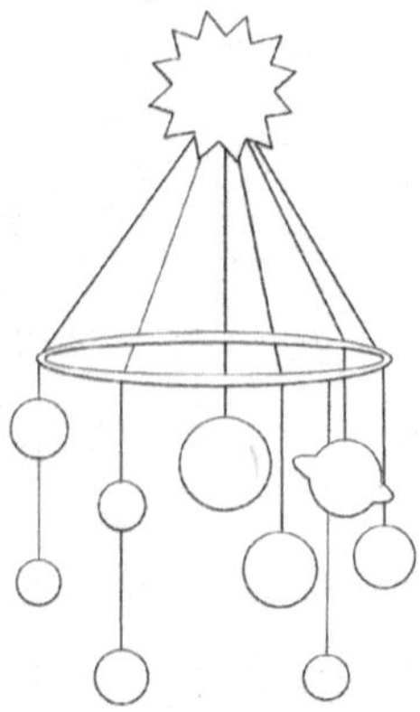

Our only refuge, fragile and immense,
a balance of chance and care.

Mars

"To Mars I call, the strongest far, unrivalled in the dreadful deeds of war."
Homeric Hymn to Ares

"Mars is the most hospitable of the planets after Earth. It has become the great frontier of our age."
Stephen Hawking

"Out of this world perhaps once more to be renewed."
John Donne

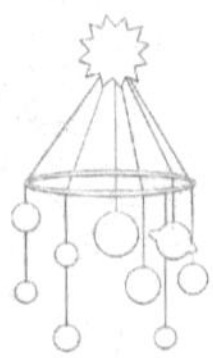

Mars, the red wanderer. To the ancients he was the god of war, his hue like blood upon the horizon, his slow march across the heavens a portent of strife. The Romans named him *Mars*, second only to Jupiter in their pantheon; the Greeks called him *Ares*, who fathered Deimos and Phobos; fear and terror themselves. We have long regarded him with awe, suspicion, and fascination.

With the telescope came intrigue: maps traced with imagined canals, the work of other minds upon another world. That dream dissolved, but the enchantment held fast. Mars became the stage for science fiction invaders and "little green men," the war-god reimagined in pulp and cinema.

In truth, Mars is no thriving world but a desert planet. Its mass is $6.42 \times 10^{\wedge 23}$ kilograms, a little more than a tenth smaller than Earth. Its atmosphere is thin, composed mainly of carbon dioxide, with a pressure of just 6.4 millibars, less than one percent of Earth's. Thin air, bitter cold, dust storms that sweep entire hemispheres. Great volcanic scars and dry riverbeds whisper of water long lost, and polar ice lies locked beneath a fragile crust.

Yet still we reach for him. Machines crawl his plains, drilling, sampling, listening. We dream of domes and colonies, of terraforming… as though the god of war might be soothed into granting sanctuary. Mars holds our contradictions: death and aspiration, ruin and promise.

We weigh the hazards and count the costs.
We gaze in wonder and dare to dream.
That this red world may one day harbour life again.

*****

Warrior

"There is a flower that grows on Mars. It is red and harsh and fit for our soil. It is called Haemanthus.
It means "blood blossom."
Pierce Brown

Red star,
eye of battle,
you rise with blood upon your shield.

Armies march to your rhythm,
empires fall in your name.

War is your colour,
war your gift,
war your curse upon us all.

Knowledge has unmasked you,
truth denying myth…
yet still we bow to your summons.

We know,
and still we fight.
We understand,
and still we kill.

*****

Sons of Ares

And thus, Eros sent Phobos & Deimos to Mars, to turn the God of War into the God of Hate.

Nicholas Chong

Phobos.
Deimos.
Fear and Dread.

Spawn of Ares,
you haunt the red god's march,
shadows chained to blood and steel.

Captured stones,
ragged, scarred,
hostages to Mars.

Phobos falls inward,
year by year,
destined for his father's embrace.

Deimos drifts outward,
watcher of loss,
the last companion left behind.

Fear.
Dread.
Forever circling.

*****

Frontier

"It might be helpful to realize, that very probably the parents of the first native born Martians are alive today."
Harrison 'Jack' Schmitt

Humankind, forever chasing
frontiers away from Eden.

We live on our jewel,
be-spoiled by dominion.

Still, we send our craft…
our rovers,
our silent scouts across your plains.

We dream of domes,
of children born beneath your sky,
of rivers remade where once they flowed.

Yet as our boots press into your dust,
what will we bring?

Life renewed?
Or death repeated?
The echo of our home?

*****

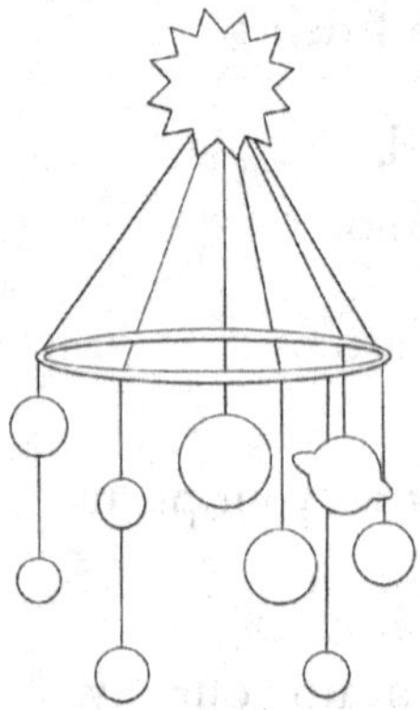

The red world waits,
a reminder of what was lost,
and what might yet be sought.

Ceres & The Asteroid Belt

"Ceres, the nourisher, gave the grain that sustains life."
Ovid, Fasti

"The asteroid belt is a failed planet; fragments left over from the Solar System's birth."
NASA

"The falling stone is each time the first."
Rainer Maria Rilke

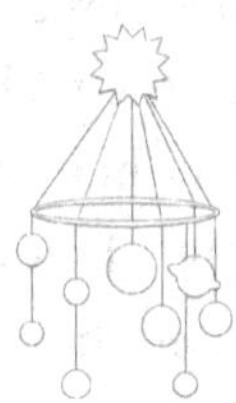

Between the rocky inner planets lies a vast belt of stone, metal, and ice… fragments of universal history, held in a fragile balance of drift and collision, of gathering and dispersal.

At its heart presides Ceres, largest of them all. Once counted as an asteroid, now named a dwarf, she reigns over this realm of broken kinship.

Too small to hold an atmosphere, she nonetheless may harbour an ocean of briny water beneath her crust, and perhaps the remnants of a molten core… conditions that hint at possibility, though the chances of life here are vanishingly slight.

How apt, then, her name: Ceres, Roman goddess of grain and harvest, protector of the plebeian class, counterpart of the Greek Demeter. From her name comes our word *cereal.* She was discovered in 1801 by Giuseppe Piazzi, the first "asteroid" to be recognised, herald of countless others caught in the Sun's pull.

The belt is the Solar System's rubble field; debris left from creation. At least thirty known worlds large enough to be named share this orbit, and doubtless more await discovery. We study them for clues to our beginnings, and we fear them for what they might become.

For the stones still wander. At any moment, one may be drawn from the dance, falling sunward, crossing our path.

Ceres reminds us of nourishment, and of ruin.
Ceres holds our origin.
Ceres holds our fate.

*****

Ceres

"She gave the grain that sustains life."
Ovid

Mother of harvest,
mistress of stone,
you reign among fragments,
a queen without throne.

No sheaves here,
no corn to bind,
only ice beneath dust,
only silence behind.

You promise abundance,
yet hold only loss.
You are seed and husk together,
our bread, and our dust.

*****

Vestigial Threat

"Noise proves nothing. Often a hen who has merely laid an egg cackles as if she laid an asteroid".
Mark Twain

O Vesta, Pallas, Europa,
Bamberga and Nemausa,
Interamnia, Davida,
Herculina, Eunomia,
Electra, Fortuna, Bellona.

Euphrosyne, Camilla,
Cybele, Thisbe, Iris too.
Hektor, Metis, Egeria,
Hygiea, Patientia,
Amphitrite, Juno, Psyche, Doris.

These are the ones I can name.
But countless more no doubt remain…
stones without a memory,
left wandering unseen.

May none of you
forget your place.
May none of you
destroy our race.

*****

Psyche

"16 Psyche may be the exposed core of an early planet, rich in metal, a relic of formation."
NASA

Psyche,
goddess of soul,
you glitter in the dark.

Iron, nickel, gold…
a motherlode beyond imagining,
more metal than Earth itself.

We dream of your riches,
of cities forged from your veins,
machines sustained by your core.

But you are not ours to plunder.
You are remnant,
memory of creation's fire.

If we take,
let us take with care.
If we build,
let us build with reverence.

For balance once broken
cannot be remade.

*****

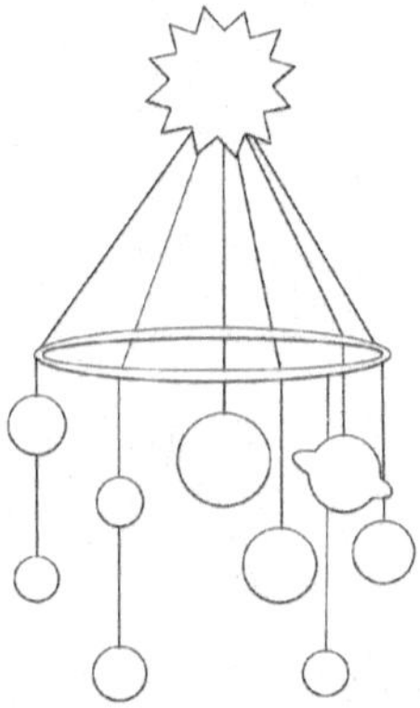

Among the fragments lies the story of beginnings,
scattered seeds across the void

Jupiter

"Father Zeus, who thunders on high, ruler of gods and men."
Homer, Iliad

"Jupiter is the great protector, sweeping up cosmic debris that might otherwise strike Earth."
NASA

"There is grandeur in this view of life – from so simple a beginning endless forms most beautiful and most wonderful have been, and are being, evolved."
Darwin

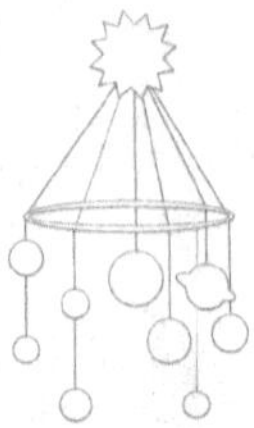

Far beyond the asteroid belt lies Jupiter. To the Romans he was king of the gods, to the Greeks, *Zeus*, all-powerful. Described as the son of Saturn and Ops, he was the sky-father, bearer of the thunderbolt, his emblem the eagle. In astrology he became a sign of abundance, expansion, and fortune.

How prescient such folklore seems.

Jupiter is eleven times the diameter of Earth, with a mass of $1.898 \times 10^{\wedge 27}$ kilograms… more than 300 times heavier than our planet, and greater than all the other worlds of the Solar System combined. Only the Sun is mightier, and Jupiter's gravity is so vast that it shapes the balance of the system itself.

On Jupiter, an object weighing 10 kilograms on Earth would weigh 25 kilograms. Yet there is no surface on which to stand: Jupiter is a gas giant, a churning ocean of hydrogen and helium with traces of methane, ammonia, hydrogen-sulphide, and water. Its atmosphere is a cauldron of storms, twisted by the Coriolis effect, with the Great Red Spot (a hurricane larger than Earth) raging for centuries.

This immense gravity well is also our protector, drawing comets and asteroids into its embrace, sparing the inner planets from endless strikes.

Jupiter's orbit around the Sun lasts 4,333 Earth days, or nearly twelve years; yet its own rotation is swift, turning once every ten hours. And Jupiter is not alone: its moons make it a system within a system. The Galilean satellites, Io, Europa, Ganymede, and Callisto; are worlds in their own right, larger than any dwarf planet. Io burns with volcanic fire; Europa hides a global ocean beneath its ice; Ganymede carries its own magnetic field; Callisto bears the scars of ancient bombardment. Each whispers possibility.

Jupiter the god, enthroned in thunder.
Jupiter the shield, protector of Earth.
Jupiter the sun-that-might-have-been.

*****

Ode to Mass

"Jupiter is more than twice as massive as all the other planets combined."
NASA

Lord of weight,
you bend the system to your will.
Eleven Earths across,
three hundred Earths in heft,
you drag the wandering stones
into your grasp.

Your gravity is anchor,
your presence, law.
You are the balance
between Sun and void,
the fulcrum of our fragile dance.

Mass is your sceptre,
power your crown.
King of the planets,
you reign unseen…
a sun unlit,
a god unbound.

*****

The Galileans

"Four worlds, each a planet in its own right."
Galileo

Io speaks:
I am fire.
I break, I boil, I burn.
Volcano is my voice.

Europa speaks:
I am silence.
I hide the sea below.
Life waits in my dark.

Ganymede speaks:
I am giant.
I carry iron in my heart.
I hold a field of my own.

Callisto speaks:
I am ancient.
I am scar and stone.
I keep the memory of strikes.

All speak:
Together we circle.
Fire, Ice, Metal, Stone.
Bound to Jupiter's will.
Worlds entire,
yet never free.

*****

The Red Spot

"A storm larger than Earth, raging for centuries."
ESA

Great mill of motion,
grinding centuries to dust,
you turn and turn,
a wheel without rest.

Eye of Jupiter,
cyclone unblinking,
you scan the Sun,
you glare at Earth,
you sweep the inner worlds
with your spiral gaze.

Scar of the sky-father,
wound that will not heal,
you are fury made visible,
eternity caught in storm.

*****

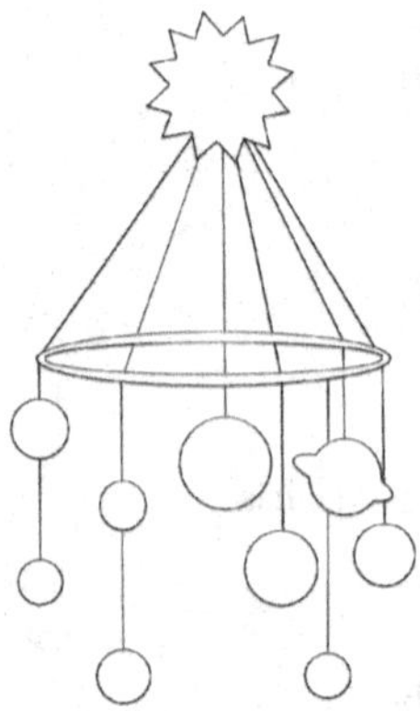

Giant of storm and mass. Protector, destroyer, and keeper of the swarm.

Saturn

"Saturn devouring his son."
Ovid, Metamorphoses

"Saturn's rings are transient, a passing show in cosmic time."
Cassini Mission Report

"Ring out, wild bells, to the wild sky."
Tennyson

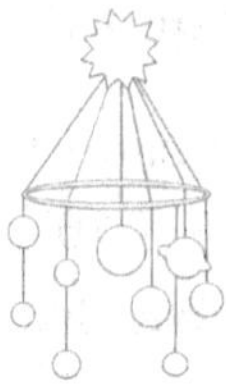

In Roman and Greek mythology, Saturnus - Cronus to the Greeks - was god of agriculture, time, and prosperity. His festival, Saturnalia, long predated Christmas, marking mid to late December when Saturn gleams cold and clear in the winter sky.

The Latin *satus* means "sowing" or "beginning": an apt root for a celebration that honoured both the close of the darkest days and the renewal to come. In myth, Saturn reigned over a Golden Age with his wife Ops, goddess of abundance. Yet fear gnawed at him: a prophecy foretold that one of his sons would usurp him. In dread, he devoured each newborn, until Ops hid Jupiter away and fate unfolded.

Artists gave him the wings of time and the scythe of harvest; a figure both generative and devouring, embodying the cycles of growth and decline.

The planet that bears his name is the Solar System's second giant, with a mass of $5.68 \times 10^{\wedge 26}$ kg, a year that spans thirty Earth years, and a day that lasts but eleven hours. Thought to possess a rocky core wrapped in metallic hydrogen, Saturn is enveloped in layers of liquid hydrogen and helium. Its outer atmosphere, tinted by drifting ammonia crystals, gives it the pale-yellow hue we see from Earth.

Saturn's grandeur lies in its retinue. At least 274 moons orbit with its vast rings of ice and dust. Among the great moons - Mimas, Enceladus, Tethys, Dione, Rhea, Hyperion, Iapetus, and Phoebe - Titan is supreme: larger than Mercury, rivalled only by Jupiter's Ganymede, with a thick nitrogen atmosphere and methane seas where alien chemistry stirs. Enceladus too is alive with promise, venting plumes of saltwater and organics into the void.

The rings, for all their beauty, are ephemeral. To human eyes they seem eternal, yet they are in slow decay, their particles raining down into the upper atmosphere to be vaporised. Born of shattered moons and ancient collisions, they will fade to nothing within 100 million years… the blink of an eye in cosmic time.

Saturn, father of the pantheon.
Saturn, keeper of time.
Saturn, counterweight among the outer worlds.

*****

Veil of Saturn

"All is flux, nothing stays still".
Heraclitus

A crown of glass and ruin,
shattered moons ground fine,
circling in silence,
a diadem of ice and dust
that glitters as though eternal
yet slips, grain by grain,
into the hunger of the sky.

Tears born of violence,
an iridescent bruise alight.
Infanticide, denied prophecy:
The veil a testament to shame,
a mask for the godly face.
Saturn, father eternal, bearing blame.

What myth foretold, the son fulfilled:
Jupiter rose, claimed the throne,
leaving Saturn to mark the hours,
to count the unravelling days.
Time's decay, flux unending…
yet for now the balance holds fast.
Still, nothing ever lasts.

*****

Chronos

"Time is the wisest counsellor of all."
Pericles

Thirty winters make a year,
eleven hours make a day,
yet still the scythe is lifted,
invisible, unerring,
a slow arithmetic of loss.

Even rings must fall,
their radiance measured out
in vanishing.

A god of measured time,
his face aligned with Earth:
winter's festival of endings,
harvest gone,
rebirth awaited.

Pay heed, the watchmaker waits.
All calibres unwind,
perpetual motion falters…
time cannot be denied.

The mainspring wound by gravity,
the balance wheel of orbits,
the jewels of cosmic birth…
all subject to decay.

Time is the only certainty.
It comes to one and all:
the ticking of millennia,
the fall of every grain.

*****

Guardians

Gravity is the music of the spheres made visible.
from Kepler

What orrery is more intricate
than the great array of Saturn?
Flocks of errant debris,
gravity-enthralled,
tended by the nine bright moons,
herdsmen of the pattern.

They pace their orbits faithfully,
silent shepherds keeping watch,
corralling fragments,
holding brightness in its place.
Each moon a note in Cronus' chord,
a resonance that steadies
the fragile veil from scattering.

Great Titan holds sway
over guardians of the field;
Enceladus, Mimas, Tethys
work their dogs to steer the horde.
Phoebe, Dione, Hyperion,
Iapetus and Rhea…
nipping at the heels of rocks,
turning comets from their course.

The flock endures,
for now in harmony;
music in orbit,
gravity made visible.

*****

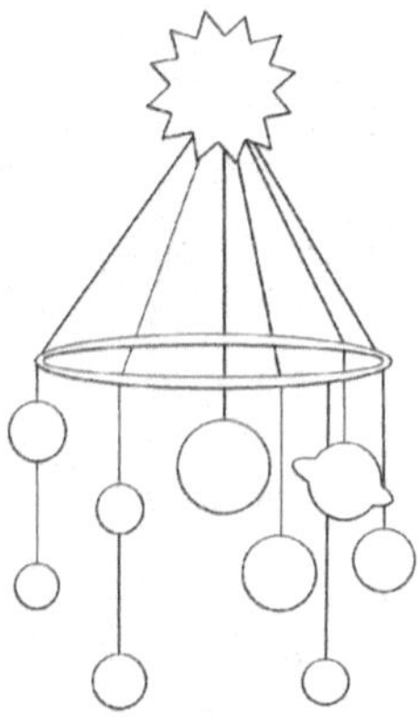

Rings of splendour, fleeting as frost
a crown that time will surely claim.

Uranus

"Uranus, the starry sky, was the first ruler."
Hesiod, Theogony

"Uranus rolls around the Sun on its side - a giant tipped by ancient collision."
NASA

"And yet the sky is not the limit."
Emily Dickinson *(adapted)*

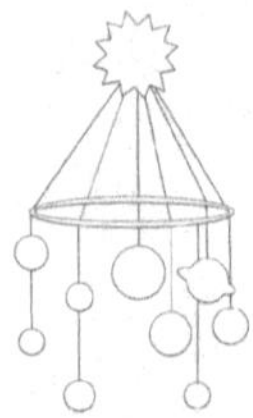

For the Greeks, Uranus – sometimes rendered *Ouranos* – was the primordial deity of the sky. He was both son and husband of Gaia (Earth), with whom he formed the foundation of the cosmos. Some tales describe him as born without a father; others say he was the child of Gaia and Aether, or of chaos itself. Uranus fathered the Titans, as well as the Cyclopes and the Hecatoncheires, the hundred-handed giants.

Unlike other ancient gods, he has few Roman parallels. His supposed counterpart was Caelus, but Uranus belongs most securely to the Greeks. Even there, he was never revered as a prime deity: less a figure of cult than a primordial force of nature.

Later astrological tradition made him the planet of disruption, revolution, and sudden insight – though the reasons for these associations are tenuous.

Astronomically, Uranus is strange. Discovered by William Herschel in 1781, it is classed as an ice giant and gleams pale blue-green, its colour shaped by methane in its atmosphere. Beneath the clouds lies a composition of water, ammonia, and methane ice, layered into a dense, opaque sphere.

With a mass of 8.68 × 10^25 kg, Uranus is the third-largest planet in the Solar System. Its most striking feature is its axial tilt: 82.23 degrees, almost on its side. Unlike most planets, it spins east to west – a trait it shares only with Venus – and the cause is still uncertain. A colossal collision in the distant past is thought to have tipped it over, a blow that may also account for its faint, narrow rings.

One Uranian year lasts 84 Earth years, while its day is just 17.5 hours. Combined with the tilt, this means each pole experiences 42 years of sunlight, followed by 42 years of darkness. In those long winters, Uranus plunges to the coldest temperatures of any planet.

Twenty-nine moons circle the planet. Five are considered major: Miranda, Ariel, Umbriel, Titania, and Oberon – names drawn, like many of the others, from Shakespeare. Fourteen inner moons cluster close to the rings, while ten irregular moons wander further out, supplying dust and fragments that feed the delicate ring system before raining down upon the planet below.

The major moons bear scars of volcanism and canyon-cutting forces, yet all remain inhospitable.

Uranus the primordial sky.
Uranus the sideways giant.
Uranus the slanted blue marble, rolling to its own rhythm.

*****

Wilful Difference

"I can do anything I want; I'm eccentric!"
John Cleese

There once was a planet askew,
That spun in a manner quite new.
It toppled and tumbled,
Its seasons all jumbled,
And puzzled the sky-watching crew.

It rolls on its side round the Sun,
As if space were a game to be won.
With moons from the plays
Of long-vanished days,
It shows that the cosmos has fun.

*****

An Alternate Path

"Stepping onto a brand-new path is difficult, but not more difficult than remaining in a situation which is not nurturing to the whole woman."
Maya Angelou

It turns where others do not care to turn,
A world of turquoise lying on its side.
Forty years of day, then night to burn,
The longest seasons any star can bide.

Some ancient blow has marked it, tipped it strange;
A crooked roll, a quiet, lonely path.
No symmetry, no tidy, ordered range,
But something scarred by history's aftermath.

And yet this slant suggests another way:
That difference may kindle thought and fire.
To move askance is not to go astray,
But live a truth that others might require.
So spins this planet, quiet, far apart,
The sideways emblem of a tilted heart.

*****

Incomplete Works

"O, swear not by the moon, the fickle moon, that monthly changes in her circle orb, lest that thy love prove likewise variable".

William Shakespeare, Romeo and Juliet *(Act 2, Scene 2)*

A company of moons takes up their parts:
Titania rules, Miranda bears her scars.
Oberon watches, Ariel departs,
Umbriel lingers, faint among the stars.

Their names are borrowed from a poet's stage,
But ice and shadow shape their only speech.
A frozen script repeats from age to age,
Beyond the playwright's ink, beyond our reach.

It seems unfinished, yet perhaps that's right:
A story that will never quite be whole.
A play of fragments, half-concealed in night,
Still whispering to the tilted planet's soul.
Not every work can tie its endings fast;
Some tales are richer for the gaps they cast.

*****

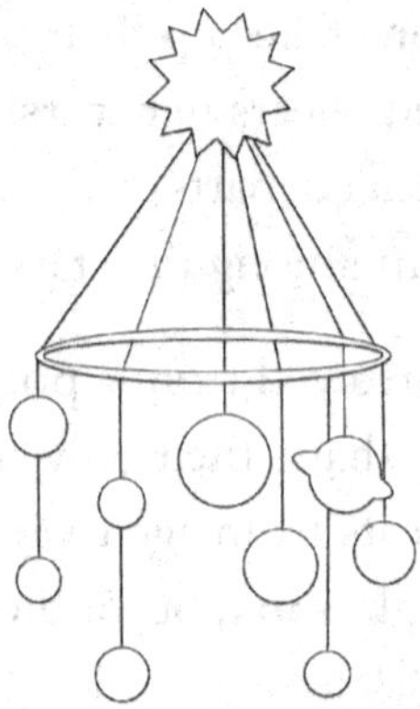

Tilted and remote, it turns askew
a quiet enigma in the outer dark.

Neptune

“O Poseidon, shaker of the earth, dark-haired god of the deep.”
Homeric Hymn to Poseidon

‘Neptune’s winds are the fastest in the Solar System, reaching supersonic speeds.”
Voyager II Report

‘Full fathom five thy father lies; of his bones are coral made.”
Shakespeare, The Tempest

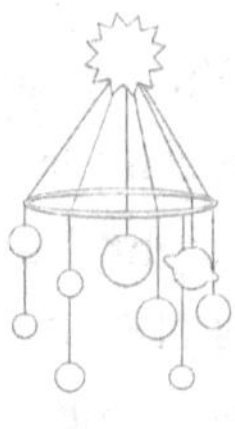

The mythology associated with Neptune is ancient and resonant. Brother to Jupiter and Pluto, he presided over the seas while they governed the heavens and the underworld.

Neptune, the Roman counterpart to Poseidon, ruled salt and fresh waters alike, alongside his consort Salacia, goddess of calm seas. In belief systems and astrology, he is linked to inspiration, illusion, and the dream world.

His festival fell in midsummer, when water was scarce and prayers rose for autumnal rains. He was one of the few deities honoured with the sacrifice of a bull, testament to his power. As *Neptunus Equestris*, he was also patron of horses and horse-racing in Rome. Scholars trace him further back still, to Proto-Indo-European water gods, with echoes in Celtic deities such as Nechtan and Nuada.

The planet itself is bound to the god only by naming convention. It holds a special place in astronomy: the first planet discovered not by sight, but by mathematics. Galileo (1612) and John Herschel (1830) both recorded it in their notes, mistaking it for a star due to its slow drift against the heavens. After William Herschel's discovery of Uranus in 1781, astronomers detected small irregularities in Uranus's orbit. Two men - Urbain Le Verrier in France and John Couch Adams in England - independently calculated that only the gravity of another, unseen planet could explain the perturbations.

In 1846, Johann Gottfried Galle, guided by Le Verrier's calculations, located Neptune from the Berlin Observatory, within a single degree of the predicted position. It was a triumph not just for astronomy but for Newtonian physics itself. Galle suggested the name *Janus*, while others proposed *Oceanus*. Le Verrier's choice, *Neptune*, won universal recognition.

Neptune has a mass of 1.0244×10^{26} kg - smaller than Uranus, yet denser and heavier. Its iron–nickel core is wrapped in silicates and a vast mantle of frozen water, ammonia, and methane. Above lies an atmosphere of hydrogen and helium, tinged with methane, which scatters blue light and gives the planet its striking hue.

Though dim from Earth, Neptune is encircled by five faint rings. It is raked by the fastest winds in the Solar System, supersonic gales that sweep across its surface. The planet hosts a colossal storm - the Great Dark Spot - echoing Jupiter's red eye.

Fourteen moons whirl around it, all named for water nymphs and sea gods. The largest, Triton, moves in retrograde orbit, suggesting it was captured by Neptune's gravity rather than born of collision.

A Neptunian year spans 165 Earth years: its days, by contrast, rush by in just 16.5 hours. This rapid spin drives the planet's wild weather and endless storms.

Neptune the deity of hidden depths.
Neptune the eternal tempest.
Neptune the anchor of our outer system.

*****

The Hidden Realm

"There are more things in heaven and earth, Horatio, than are dreamt of in your philosophy."
Shakespeare, Hamlet*, Act 1, Scene 5*

There is a kingdom no sunlight enters,
where silence is thicker than stone,
and pressure folds the world in on itself
until only darkness can endure.

Yet even here, in the farthest reaches,
there are glimmers,
not of flame or daylight,
but of secrets becoming visible for a moment,
like dreams rising and falling
in the sleep of the ocean.

This is the realm of Neptune:
not the trident-wielder of story,
not the roaring breaker on a summer shore,
but the patient sovereign of depth,
who stirs whole continents with the brush of his hand,
and bends the paths of worlds
with the gravity of his silence.

He is illusion and inspiration,
mirage and memory;
he speaks not in thunder,
but in the slow migration of currents,
in tides that climb the moon's invisible stair,
in storms that circle for centuries unseen.

To enter his dominion
is to surrender certainty;
for here, what is hidden is stronger than what is shown,
and the unseen governs the visible,
as surely as the dream shapes the waking day.

*****

The Outer Well

"Gravity explains the motions of the planets, but it cannot explain who sets the planets in motion."
Sir Isaac Newton

From distant dark, a well invisible,
draws silently the wanderers in their turn;
a hand unseen, yet infinitely full,
that bends the paths of comets as they burn.
The further edge of what is knowable,
its blue abyss where latent forces churn,
a watchful depth, remote, inscrutable,
whose presence every circling world must learn.

Yet at the turn, the system holds in place…
a dance of balance, endlessly precise,
the outer anchor binding inner grace,
orbital music, measured, cold, concise.
What seems disorder proves a hidden art:
one centre, drawing all things to its heart.

*****

The Calculus of Balance

"The orbit of Uranus could not be reconciled with theory…
until Neptune was found, just where mathematics foretold."
***Cambridge Observatory Notes**, 1846*

The orbit strayed, a fraction out of line,
a whisper threaded through celestial spheres.
Each number tested, tightened to refine,
the calculus repeating all its tiers.

A whisper threaded through celestial spheres,
an unseen weight that tugged a path aside.
The calculus repeating all its tiers,
returned, revised, until the sums complied.

An unseen weight that tugged a path aside,
a presence hidden, yet its hand was sure.
Returned, revised, until the sums complied,
and proof emerged the mind could now secure.

A presence hidden, yet its hand was sure…
the orbit strayed, a fraction out of line.
And proof emerged the mind could now secure:
Neptune revealed, where numbers made it shine.

*****

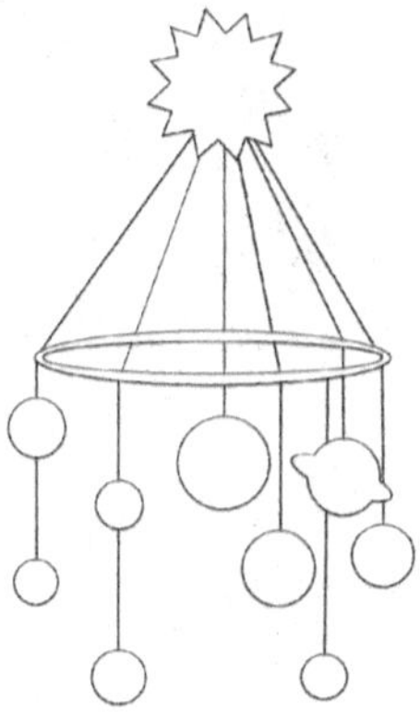

Lord of winds, deep blue and restless,
his gaze fixed ever outward.

Pluto

"Pluto, lord of the underworld, unseen by mortal eyes."
Virgil, Aeneid

"Pluto is more active than expected: a world of nitrogen ice plains and hidden oceans."
New Horizons Report, 2015

"Call no man happy until he is dead."
Sophocles

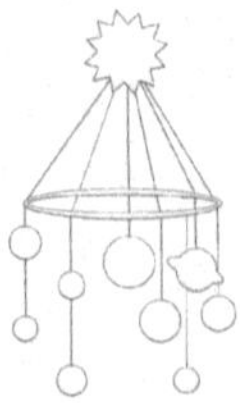

Pluto, the Roman god of death and the underworld, was equivalent to the Greek Hades, guardian of souls in their passage to the underworld. His name derives from *Ploutos*, "wealth," evoking the mineral riches of the earth beneath.

He abducted Proserpina, daughter of Ceres, whose grief gave rise to the seasons. In later astrology, Pluto came to symbolise transformation and renewal, a stern guide through shadow and trauma.

Astronomically, Pluto was discovered in 1930 by Clyde Tombaugh of the Lowell Observatory, following Percival Lowell's theory of a "Planet X" tugging at Uranus and Neptune. Using blink comparison, Tombaugh identified the faint wanderer. The name was suggested by Venetia Burney, an eleven-year-old schoolgirl in Oxford, whose idea her grandfather passed to Lowell Observatory.

Yet Pluto was too small to fit the grand theory. Though accompanied by its moon Charon, it lacked the mass to clear its orbital path. In 2006 the International Astronomical Union reclassified Pluto as a dwarf planet. Planet X remained a phantom.

Then, in 2015, New Horizons swept past. Far from a frozen husk, Pluto was revealed as a world of nitrogen ice plains, mountains of water ice, and a thin, shifting atmosphere. Cryovolcanoes and fractures hint at a hidden ocean beneath the crust. With a mass of 1.31×10^{22} kg – just 17% that of our Moon, and comparable in size to Antarctica – Pluto proved itself active, enigmatic, and alive with story.

Pluto, Lord of the Dead.
Pluto, diminished yet defiant.
Pluto, the outlying mystery.

*****

Elegy for Lost Status

"Pluto was a planet for 76 years. That's a lot of homework."

Neil deGrasse Tyson

Once we learned them, one through nine,
A classroom litany we'd recall.
Your name intoned in ordered line,
The smallest sovereign of them all.

But numbers told a colder truth:
Your mass too slight, your power small;
Dethroned, denied your "X," your proof,
Uncounted from the planet-haul.

Yet still we mourn with wistful eyes,
Upon your plains of frozen gas;
A ghostly world, where silence lies,
An exile drifting, yet steadfast.

*****

Defiance

I am that which is, and was, and shall be, and no mortal has ever lifted my veil.

from an inscription, Temple of Isis

You call me "dwarf," diminished, redefined,
Yet still I circle, steadfast, undeterred;
Your fleeting labels cannot make me mind.

I quake, my mountains rise, my heart is stirred;
Beneath my crust, a secret ocean lies,
A hidden story, no council ever heard.

I am the underworld, where silence cries;
Uncrowned, I stand, defiant in the cold,
My story written deep in starless skies.

Forget your rules, your kingdoms weighed in gold:
Beyond your maps, my watchful orbit keeps,
Enduring, patient, dignified, and bold.

*****

Invitation

And beyond the known, the unknown draws us still.
Anonymous, ancient seafarer's proverb

A frozen scree beyond my watch extends,
Where countless bodies drift in icy seas;
The journey outward never truly ends.

A million wanderers, faint and dim, are friends,
They circle slow, with mysteries to tease;
A frozen scree beyond my throne extends.

The giants tilt, their stately orbit bends,
Some unseen weight still whispers through the freeze;
The journey outward never truly ends.

Perhaps a monarch hidden there attends,
A shadow planet veiled in dark decrees;
A frozen scree beyond my throne extends.

So, traveller, go on where silence lends,
Through Kuiper's belt, beyond my boundaries;
The journey outward never truly ends.

*****

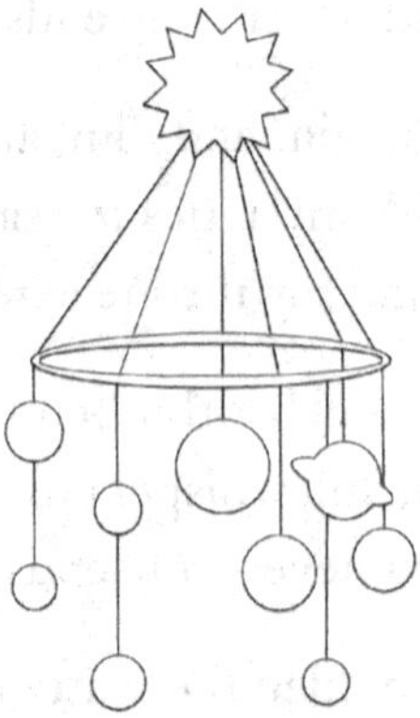

Pluto stands as threshold, neither lost nor found, but waiting, the last sentinel before the sea of countless worlds.

The Kuiper Belt

"Eris, strife incarnate, delights in discord."
Hesiod, Theogony

"The Kuiper Belt is a frozen frontier… countless small worlds beyond Neptune's sway."
NASA

"The edge of things… where maps must end."
Seamus Heaney

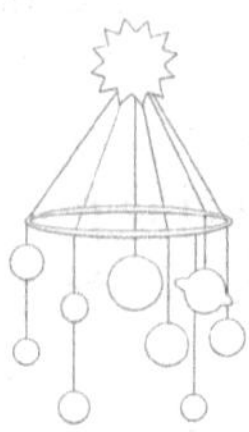

This far out in the Solar System, our inherited mythologies fall silent. Beyond Neptune, the gods of Olympus and Rome no longer hold sway. Here, in a realm without ancient stories, we begin to write new ones of ice and shadow.

Pluto's discovery in 1930 offered the first hint of a hidden population, though it was only in 1951 that astronomer Gerard Kuiper proposed the existence of a distant belt of matter. Even then, his idea remained conjecture until 1992, when the first Kuiper Belt Object (KBO) was confirmed.

What we now call the Kuiper Belt is thought to be an immense structure: A tenuous ring orbiting the Sun beyond the planets, akin to the Asteroid Belt but vastly greater in scale, perhaps between 100 and 200 times more massive. (NASA)

Its composition is still being pieced together. The belt appears to consist mainly of frozen volatiles (methane, ammonia, water ice) locked in countless small bodies. Mixed among them may be rocky remnants from the Solar System's birth, fragments that never accreted into full worlds.

Among this drifting multitude, a handful of larger objects have been identified: Pluto, Haumea, Makemake, Eris, Arrokoth. Each adds nuance to our understanding. Eris, slightly larger than Pluto and attended by its moon Dysnomia, revealed that "planet X" was not a single hidden world but an entire system within a system.

Yet these are only the first glimpses. Astronomers estimate there may be hundreds, perhaps hundreds of thousands, of KBOs more than 100 kilometres across, most still unseen. It is now thought that many short-period comets, with orbits of less than two centuries, originate in this frozen frontier.

The Kuiper Belt remains a boundary of speculation as much as of knowledge. Its icy bodies may preserve primordial matter from the dawn of the Solar System, their chemistry suspended in deep cold, awaiting discovery. But for now, much of what lies beyond Neptune is inferred, not known.

Kuiper Belt, frontier of our knowledge.
Kuiper Belt, keeper of secrets.
Kuiper Belt, home to the new gods.

*****

Outer Chorus

"All the world's a stage, and we are merely players, performing in a grand choir of life."
William Shakespeare, As You Like It, *Act 2, Scene 7*

Far off in the outer reaches
Circling the sun.
We're remnants of your building blocks
Around what you've become.

Pluto
You named me for imagined heft
But you misunderstood.
I'm just another planetoid
And doing what I should.

Far off in the outer reaches
Circling the sun/
We're remnants of your building blocks
Around what you've become/

Haumea
How strange it is to be thought as
a Hawaiian god of yore.
More barren than fecund, you see?
Your irony is poor.

Far off in the outer reaches
Circling the sun.
We're remnants of your building blocks
Around what you've become.

Makemake
An Easter Island demigod
Born of creation myth
The link is rather tenuous
But it's what I'm lumbered with.

Far off in the outer reaches
Circling the sun.
We're remnants of your building blocks
Around what you've become.

Eris
Now just a crucial minute!
(interruption, as ever mine)
You found *him* first, but greater still
Is discord: I'm Eris, most divine.

Far off in the outer reaches
Circling the sun.
We're remnants of your building blocks
Around what you've become.

Arrokoth
The god of sky for Powhatan tribes
My lobes are Wenu Weeyo…
I bide my time, observant, proscribed
Our kin may rain upon you…

Far off in the outer reaches
Circling the sun.
We're remnants of your building blocks
Around what you've become.

All KBO's
We hope that curiosity quells
The ignorance you fought.
We KBO's more numerous
Than the numbers that you've wrought.

Far off in the outer reaches
Circling the sun.
We're remnants of your building blocks
Around what you've become.

****

Building Blocks

'Most of the time in the 21st century, we dominate our surroundings: We tweak the thermostat, and the temperature falls one degree. We push a button, and Taylor Swift sings for us. It's the opposite in the wilderness, which teaches us constantly that we are not lords of the universe but rather building blocks of it."
Nicholas Kristof

From silence scattered, all begins as stardust,
A drifting veil that thickens into matter,
Cold vapour crystallised as first bright water,
Strange pull of gravity becoming bond,
Infinitesimal motes grown dark with minerals,
A slow ballet of newly named elements.

Each spark of order from chaotic elements
Coalesces, faint and shimmering with stardust,
Until attraction stitches mass from minerals,
A congregation of what we call matter.
The tug is constant, binding force to bond,
Linking earth and sky with seed of water.

Upon the young world fell that gift of water,
Bringing breath to stone and breath to elements;
A fragile covenant, a planetary bond
First written in the ashes of stardust.
Yet we forget the finite weight of matter,
Hewing deep, extracting living minerals.

We dig rapacious seams of rare minerals,
Drain aquifers and squander ancient water,
Forget the delicate arithmetic of matter,
Or how we are no greater than our elements,
All woven from the same essential stardust,
Each covenant a thread in nature's bond.

Hubris frays the fabric of the bond,
Profligacy dissolves in poisoned water;
We strip the soil of irreplaceable minerals,
Forget that all our blood is born of stardust,
That all our breath is borrowed from these elements,
That even kings are crumbled back to matter.

So, turn again towards the roots of matter,
Restore the covenant, remember the bond;
Live humbly with the balance of the elements,
Protect the fragile arteries of water,
Tread lightly on the seams of buried minerals,
And know: our fate is written still in stardust.

Until we match stardust with our matter,
Cleanse our water, honour each bond,
And guard the minerals and elements.

*****

The Sky is Falling

"The sky is falling. The sky is falling; and, I fear you won't believe it until you're covered in stars"
Emalynne Wilder

The sky is falling from the frozen rim,
Oumuamua streaked and slipped away;
The orbit tilts, and worlds collide again.

Atlas broke apart, its fragments swim,
Comets dissolve in sunlight's fierce decay;
The sky is falling from the frozen rim.

Our whole system in delicate balance,
A Jenga tower of stone in fragile chain;
The orbit tilts, and worlds collide again.

Gravity's whispers, silence and violence,
Momentum's pull no mortal hand can stay;
The sky is falling from the frozen rim.

A slender margin, nothing more than chance,
Protects our world from chaos and its sway;
The orbit tilts, and worlds collide again.

No map can tame the void, no hymn restrain:
Though brief, our peace still trembles in display.
The sky is falling from the frozen rim,
The orbit tilts, and worlds collide again.

*****

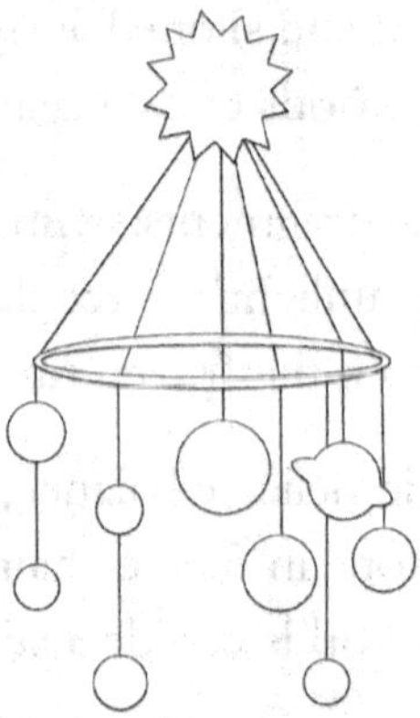

Often, the most important details
are the furthest from our grasp.

Which is why we must keep reaching.

Planet X

"Beware lest you lose the substance by grasping at the shadow."
Aesop

"The unseen moves the seen."
Tao Te Ching

"Something out there is tugging on the orbits."
Mike Brown, Caltech, 2016

"There are more things in heaven and earth, Horatio, than are dreamt of in your philosophy."
Shakespeare, Hamlet

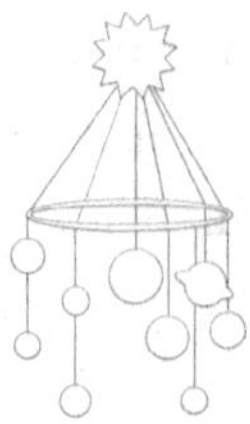

For centuries, astronomers have looked beyond the known planets and wondered what else might wait in the dark. First came speculation about Uranus's wanderings, which led to Neptune. Later, Pluto was hailed as the long-sought "Planet X," though its mass proved too meagre to explain the anomalies in the outer system. And still the question lingered: is there another world, larger, more distant, drawing subtle patterns into the orbits we can see?

The Kuiper Belt, once assumed a scattered ring, reveals strange alignments. A handful of its most elongated members - Sedna, 2012 VP113, and others - cluster together as though shepherded by some vast, unseen presence. Their orbits tilt and stretch in ways gravity alone should smooth away. The odds of such arrangement by chance are low. Something exerts a hidden hand.

In 2016, astronomers Konstantin Batygin and Mike Brown advanced the boldest modern hypothesis: a planet perhaps ten times the mass of Earth, tracing a slow, eccentric path hundreds of times farther from the Sun than our own. Not a gas giant like Jupiter, nor an ice dwarf like Pluto, but a new category altogether: a frozen super-Earth haunting the edge of the Solar System.

Others caution against premature coronations. Perhaps these alignments are illusions, artefacts of limited data. Perhaps there is no monarch in the darkness, only the echoes of chance. Yet the anomalies persist, tugging at models, whispering of more to come.

If such a planet exists, it reshapes our story of the Solar System. It may be an exile, cast outward in the early turmoil of planetary migration, or a wanderer captured from another star. It may be a phantom that explains not only the clustering of KBOs, but why the Kuiper Belt exists in its current form at all.

For now, Planet X remains unseen: a rumour written into orbital paths, a shadow tugging at the edge of maps.

*****

The Phantom Monarch

If Planet Nine exists, it has shepherded these objects into place."

Konstantin Batygin

They call me phantom, monarch without throne,
A rumour traced in others' bending ways;
Though I am unseen, their motions are my own,
A court that bows in silence to my sway.

Sedna inclines, compelled by what I bear,
The Kuiper Belt aligns as if in prayer;
A hand unmeasured holds their orbits there,
Though mortal eyes find only empty air.

Exiled perhaps, flung outward long ago,
Or captive stolen from another sun;
My kingdom lies where human charts won't go,
A slow dominion, centuries begun.

Not crown of flame, nor crown of ice I wear:
But gravitas, the weight of what's not there.

*****

Chorus of the Exiles

'We are defined not only by where we are, but by what we orbit."
Rebecca Elson

Sedna
My path is long, a thousand winters wide,
I turn, but never close the circle whole.

2012 VP113
I lean the same way, century on century,
my scarlet light too faint for human eyes.

Quaoar
I dance askew, off-kilter in the dark,
drawn by a summons I cannot name.

Leleakuhonua
A wanderer, my orbit stretched like song,
I travel further than your maps can reach.

All
Something compels us,
a pull beyond Neptune,
a gravity unseen.

Sedna
I would be lone, but others tilt as I…
a pattern carved in frost and time.

2015 TG387 (The Goblin)
My arc returns beyond imagining,
yet still I bend toward mystery.

All
Not chance, not silence,
but a hidden weight,
an unseen shepherd in the night.

*****

The Unseen Hand

Invisible threads are the strongest ties."
Friedrich Nietzsche

An unseen hand still pulls the outer skies,
Sedna inclines, compelled though none appear;
The orbits tilt, and whisper what is near.

A phantom planet hides from searching eyes,
Its weight is felt in curves that re-ensphere;
An unseen hand still pulls the outer skies.

The Goblin bends, Leleakuhonua flies,
Their distant arcs converge as if in fear;
The orbits tilt, and whisper what is near.

No crown of ice, no giant's blaze denies,
Yet quiet gravitas is written clear:
An unseen hand still pulls the outer skies.

Perhaps by chance, perhaps by fate's disguise,
But clustered paths repeat, year after year;
The orbits tilt, and whisper what is near.

So, questions haunt where no reply complies…
The dark may rule, though absent to the seer;
An unseen hand still pulls the outer skies,
The orbits tilt, and whisper what is near.

*****

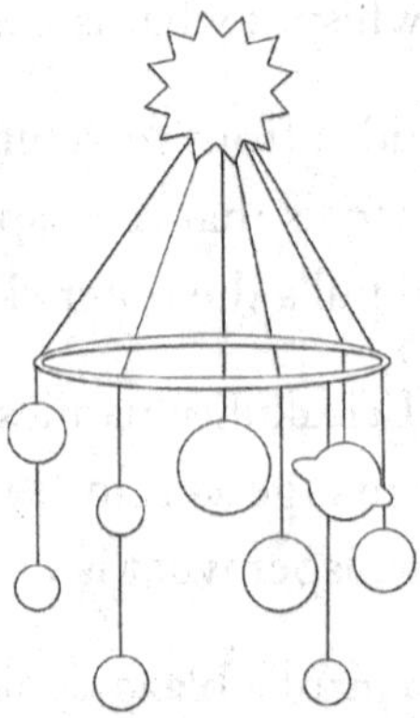

It is often the shadows and not the obvious, that teach us where to look; absence leaves the sharpest trace.

Epilogue - The Void Beyond

"In the beginning was chaos, the yawning void."
Hesiod, Theogony

"Most of the universe is not matter at all, but dark energy and dark matter… the invisible architecture of reality."
ESA/CERN

"I have loved the stars too fondly to be fearful of the night."
Sarah Williams, The Old Astronomer

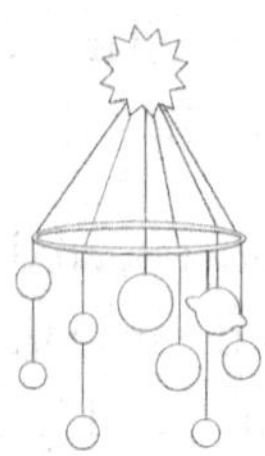

In most mythologies and cultures, there exists an idea of a darkness to be feared, an emptiness at the edge of all things. For the Greeks it was Nyx, daughter of Chaos and consort of Erebus, mother to Aether and Hemera, dwelling in the deepest recess of the underworld. Homer wrote that even Zeus himself shrank from her displeasure.

The sky has long been cast as destiny: Heaven and Hell, Paradise and Armageddon, order and chaos. Yet beyond the Kuiper Belt we step into the tenuous extremes of our knowledge.

On September 5th, 1977, NASA launched Voyager 1. Its purpose was to study the outer Solar System and whatever lay beyond the Sun's heliosphere. Almost thirty-five years later, the probe crossed that threshold; a liminal zone known as the *termination shock*, where the solar wind weakens against the interstellar medium. Nearly fifty years after launch, Voyager reached the Heliopause.

Here, scientists found unexpected heat (up to 50,000 K) though the probe slipped through unharmed, for the particles are spread unimaginably thin. This heliopause marks the edge of a great bubble: a membrane in the shape of a flattened croissant, enclosing the entire Solar System. A wall of fire, where solar wind meets the winds of other stars; like cross-currents breaking the surface of a sea.

Beyond it lies silence.

And further still, perhaps, the Oort Cloud: a vast sphere of icy planetesimals proposed by the Dutch astronomer Jan Oort. Billions of frozen fragments, too far and faint to see, yet thought to supply the long-period comets that plunge toward the Sun before vanishing into its gravity well.

The distances are immeasurable. Even at its relentless speed, Voyager will not reach the Oort Cloud for many centuries, and by then it will almost certainly be dark and mute.

We can only speculate whether other star systems are likewise wrapped in membranes, whether other Oort-like clouds wait unseen in their shadows. For now, this realm remains what it has always been: a desert of silence, lying beyond our final frontier.

*****

Terminal Silence

"The eternal silence of these infinite spaces terrifies me."
Blaise Pascal

Where the solar wind dissolves in flame,
The boundary holds, yet nothing speaks.
No signal carries a human name,
Only the hush that eternity keeps.

*****

Voyager's Lament

"Look again at that dot. That's here. That's home. That's us."

Carl Sagan, Pale Blue Dot

I will not live
to hear the probe's faint heart
reach the Oort's first frozen stones.
Three centuries of silence
stand between us.
My life, a brief inhalation;
its journey, an exhaled sigh
across the dark.

*****

The Unreachable

"So far as I can see, nothing is known."
Xenophanes

We cannot cross the void with mortal breath,
The distances outlast our fragile years,
The silence holds more firmly than our death.

We launch our machines, defying time and death,
Yet still they fall to silence with the years:
We cannot cross the void with mortal breath.

The stars look on, unmoved by birth or death,
Their light arrives through silence, dust, and years.
The silence holds more firmly than our death.

So dream we must, though silence seals our death;
Our finite span is shorter than the years.
We cannot cross the void with mortal breath,
The silence holds more firmly than our death.

*****

And Now for Something Completely Cosmic

"Life's a piece of shit, when you look at it."
Monty Python, 'Always Look on the Bright Side of Life'

Beyond the heliopause?
Nothing.
Beyond the Oort Cloud?
More nothing.
Billions of kilometres of…
well, yes…
nothing.

And here we are,
bags of water and carbon,
arguing about biscuits,
the colour of shirts and flags,
who gets to cross a line on a map,
whose money is real,
whose skin is more valued.

Meanwhile we hurl tinfoil boxes
with radios bolted on,
as if the cosmos were waiting
for our mixtape.

Three hundred years from now
Voyager might bump into
a lump of ice.

By then our descendants
will still be
trading insults online,
pricing bottled water,
and debating whether
pineapple belongs on pizza.

So what lies beyond?
Probably nothing we'll ever see.
And if Nyx herself turned up,
we'd just ask her
to put the kettle on.

*****

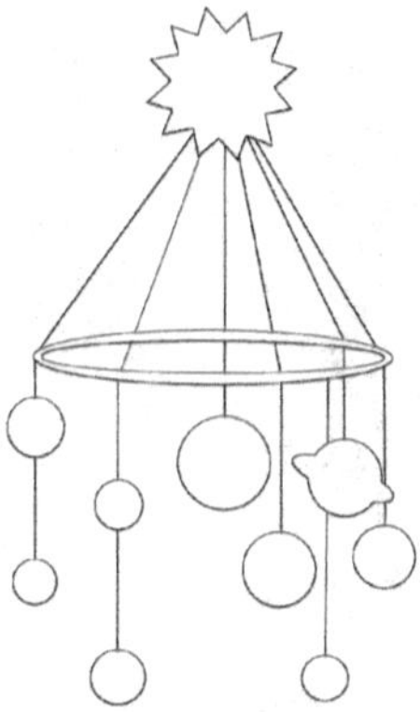

Humanity has always striven
towards knowledge over myth.
But there are limits.

Coda

"The universe is not required to be in perfect harmony with human ambition."
Carl Sagan

"What we call the beginning is often the end. And to make an end is to make a beginning."
T.S. Eliot

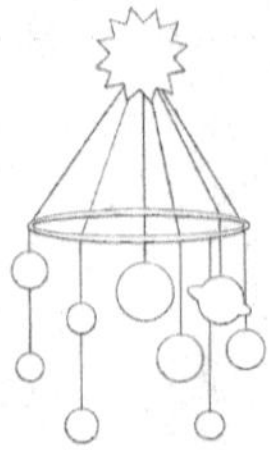

Equipoise

"Trying to save ecosystems has more to do with changing egosystems".
Don Rittner

The magnificence of Earth's entwining whole
Each interface and balance speak of care,
A fragile wisdom deep within its role.

Yet how we squander what was never spare,
Our hunger deaf to consequence or cost,
A plunder fed by greed we cannot bear.

Like fish that thrash when riverbeds are lost,
Still chasing prey, still fighting kin for more,
We squander time until the line is crossed.

Our place is thin, our foothold never sure,
Yet outward, wider systems call us still,
A harmony that waits beyond the door.

Sustaining balance rests upon our will,
To see the whole, not fragment, not the part;
One turning point, and all falls dark and still.

The future hangs on what we choose to start,
Not on ambition's grasp, but in the art
Of tending Earth, the axis of the heart.

*****

Stardust

"It's enough for me to be sure that you and I exist at this moment."

Gabriel García Márquez

What is age, if not a number?
Time, a whisper of decay
The slow unravelling of structure,
Dust that formed us, swept away.

Yet we are made of ancient starlight,
A universal, burning truth;
And holding fast to that within me
Brings me comfort, strength, and proof.

So long as sparks still cross my synapses
So long as wit and pen remain,
I'll marvel at this brief existence
Then rise to stardust once again.

*****

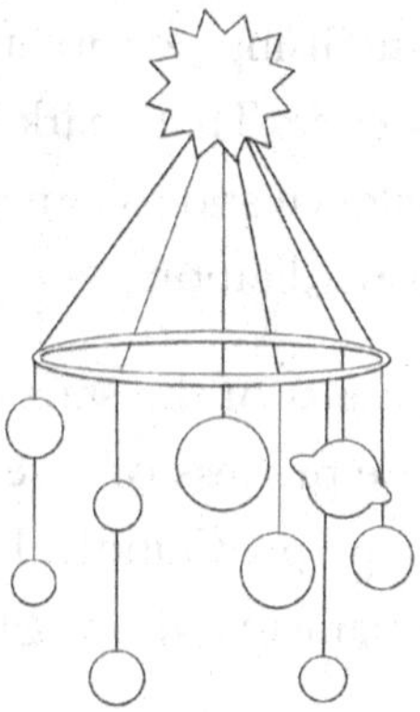

Brief yet luminous, we live in balance; every ending a beginning, all things returning to light

Afterword

"Always walk through life as if you have something new to learn and you will."
Vernon Howard

The idea for this book began with a news story about Planet X, and the tantalising possibility of yet more evidence for its existence. That spark led me to write a brief poem about balance and assumed permanence, and it grew into this larger meditation.

I already knew fragments of the science and mythology gathered here, but the process of research and writing has stretched that envelope profoundly. I now carry a deeper insight into the mechanisms of our galactic home, and with it, a deeper awe.

Yet awe is not the only feeling. Alongside it sits a sorrow: a recognition of the futility of my species and the scale of our impact on our fragile world. None of us will find a home in the stars. Our wilful ignorance, our greed, our sheer density of population, they remind me, uncomfortably, of *Vermin.*

Still, I remain hopeful. If enough of us can lift our blinkers, open our minds, and become properly curious again, perhaps we can begin to repair some of the damage. For all our cleverness, it would be an impossible task to conjure a new supply of air and water if we fail to mend what we have already broken.

What began as a stretched metaphor, of systems in balance, like a mobile, until they are not, has become something more: an environmental polemic, carried on the strange currents of mythology, science, prose, and poetry.

If any part of this work has touched or inspired you, if it has pricked a small corner of your conscience to engage differently with your fellow travellers, then I can rest assured: I have done my job.

Thank you for reading this far.

Christopher Cotton
Suffolk, 2025

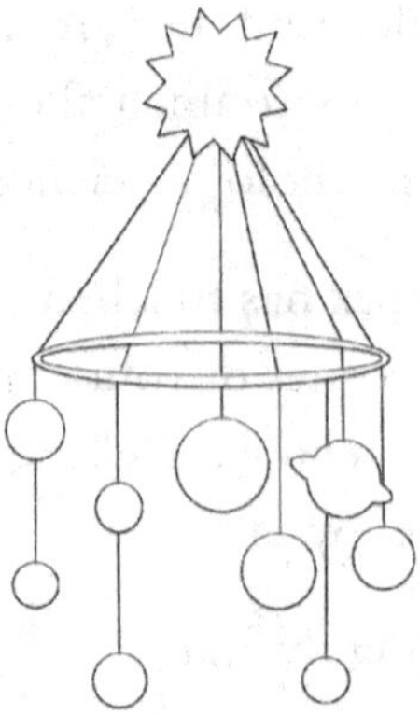

"Do the best you can until you know better.
Then when you know better, do better."
Maya Angelou

Appendix:

1) A sense of scale

The table below shows how much time it could take to travel by spaceship to various destinations in our solar system.

Destination	*Time Taken (days)*	*Time Taken (years)*
Mercury	2555	7
Venus	155	0.43
Moon	3	0.01
Mars	186	0.51
Ceres	730	2
Jupiter	1825	5
Saturn	2555	7
Uranus	3285	9
Neptune	4380	12
Pluto	5475	15
Kuiper Belt	16425	45
Planet X	18250	50

2) Attributes

Sol (The Sun)

- Mass: 1.989×10^{30} kg ($333{,}000 \times$ Earth)
- Diameter: 1.39 million km
- Composition: Hydrogen & helium plasma
- Source of solar wind & energy for the system

Mercury

- Distance from Sun: 0.39 AU (57.9 million km)
- Mass: 3.3×10^{23} kg (5.5% of Earth)
- Year: 88 Earth days
- Extreme temperature swings: +430 °C to –180 °C

Venus

- Distance: 0.72 AU (108 million km)
- Mass: 4.87×10^{24} kg (81.5% of Earth)
- Atmosphere: CO_2, thick clouds of sulphuric acid
- Surface pressure: $90 \times$ Earth; hot enough to melt lead

Earth

- Distance: 1 AU (150 million km)
- Mass: 5.97 × 10^^{24} kg
- Atmosphere: N_2 & O_2, sustaining life
- Only known inhabited world

Mars

- Distance: 1.52 AU
- Mass: 6.42 × 10^^{23} kg (10.7% of Earth)
- Thin CO_2 atmosphere; giant volcanoes & dust storms
- Polar ice caps, evidence of ancient water

Ceres & The Asteroid Belt

- Ceres mass: 9.4 × 10^^{20} kg (dwarf planet)
- Asteroid Belt lies between 2.1–3.3 AU
- Fragments from Solar System formation
- Possible subsurface brine in Ceres

Jupiter

- Distance: 5.2 AU
- Mass: 1.898×10^{27} kg ($318 \times$ Earth)
- Gas giant, no solid surface
- Over 90 moons, including Io, Europa, Ganymede, Callisto
- Great Red Spot: storm >300 years old

Saturn

- Distance: 9.5 AU
- Mass: 5.68×10^{26} kg
- Famous ring system of ice & dust (temporary)
- Largest moon: Titan (thick atmosphere, methane seas)
- 146 confirmed moons (2025)

Uranus

- Distance: 19.2 AU
- Mass: 8.68×10^{25} kg
- Ice giant tipped on its side (axial tilt 98°)
- At least 27 moons, many with Shakespearean names

Neptune

- Distance: 30 AU
- Mass: 1.02 × 10^26 kg
- Strongest winds in the Solar System (up to 2,100 km/h)
- Largest moon: Triton (retrograde orbit, likely captured)

Pluto

- Distance: ~39 AU (varies)
- Mass: 1.31 × 10^22 kg
- Dwarf planet with thin nitrogen atmosphere
- Largest moon: Charon; evidence of subsurface ocean

Kuiper Belt

- Region: ~30–50 AU
- Contains Pluto, Haumea, Makemake, Eris, Arrokoth
- Source of many short-period comets
- Vast population of icy bodies and remnants

Planet X (Hypothetical)

- Proposed mass: ~10 × Earth
- Possible orbit: 400–800 AU
- Evidence: clustering of trans-Neptunian object orbits
- Not yet observed

*****

3) Epigraphs

Foreword

- "If you look at Earth from space you see a dot…" ***Carl Sagan***
- "The fact that we live at the bottom of a deep gravity well…" ***Douglas Adams***
- "Wheels within wheels in a spiral array…" ***Neil Peart***

Prologue

- "In the beginning he established the balance…" ***Enuma Elish***
- "The fault, dear Brutus, is not in our stars…" ***Shakespeare***
- "Equilibrium is the profoundest tendency of all matter." ***Goethe***

Sol

- "Helios, lord of the sun…" ***Homeric Hymn***
- "The sun… can still ripen a bunch of grapes…" ***Galileo***
- "Even after all this time, the sun never says to the Earth…" ***Hāfez***

<u>**Mercury**</u>

- "Hermes, messenger of the gods…" ***Homeric Hymn***
- "Mercury's orbit is the most capricious of all…" ***James Jeans***
- "Fleet-footed Hermes, bringer of dreams…" ***Homer***

<u>**Venus**</u>

- "Venus herself, with all her beauty…" ***Ovid***
- "Venus is a hell-world…" ***Carl Sagan***
- "She walks in beauty, like the night…" ***Byron***

<u>**Earth**</u>

- "Gaia, mother of all…" ***Homeric Hymn***
- "The Earth is the only world known to harbour life…" ***Carl Sagan***
- "The Earth is what we all have in common." ***Wendell Berry***

Mars

- "To Mars I call…" ***Homeric Hymn***
- "Mars is the most hospitable of the planets…" ***Stephen Hawking***
- "Out of this world perhaps once more to be renewed." ***John Donne***

Ceres & Asteroid Belt

- "Ceres, the nourisher…" ***Ovid***
- "The asteroid belt is a failed planet…" ***NASA/JPL***
- "The falling stone is each time the first." ***Rilke***

Jupiter

- "Father Zeus, who thunders on high…" ***Homer***
- "Jupiter is the great protector…" ***NASA***
- "There is grandeur in this view of life…" ***Darwin***

Saturn

- "Saturn devouring his son." ***Ovid***
- "Saturn's rings are transient…" ***Cassini Mission***
- "Ring out, wild bells, to the wild sky." ***Tennyson***

Uranus

- "Uranus, the starry sky…" ***Hesiod***
- "Uranus rolls around the Sun on its side…" ***NASA***
- "And yet the sky is not the limit." ***Emily Dickinson (adapted)***

Neptune

- "O Poseidon, shaker of the earth…" ***Homeric Hymn***
- "Neptune's winds are the fastest…" ***Voyager II Report***
- "Full fathom five thy father lies…" ***Shakespeare***

Pluto

- "Pluto, lord of the underworld…" ***Virgil***
- "Pluto is more active than expected…" ***New Horizons Report***
- "Call no man happy until he is dead." ***Sophocles***

Kuiper Belt

- "Eris, strife incarnate…" ***Hesiod***
- "The Kuiper Belt is a frozen frontier…" ***NASA***
- "The edge of things…" **Seamus Heaney**

Planet X

- "Beware lest you lose the substance…" ***Aesop***
- "The unseen moves the seen." ***Tao Te Ching***
- "Something out there is tugging…" ***Mike Brown***
- "There are more things in heaven and earth, Horatio…" ***Shakespeare***

Epilogue

- "In the beginning was chaos…" ***Hesiod***
- "Most of the universe is not matter at all…" ***ESA/CERN***
- "I have loved the stars too fondly…" ***Sarah Williams***

Coda

- "The universe is not required…" ***Carl Sagan***
- "What we call the beginning is often the end." ***T.S. Eliot***

Afterword

- "Always walk through life as if you have something new to learn…" ***Vernon Howard***

- "Do the best you can until you know better…" ***Maya Angelou***

*****

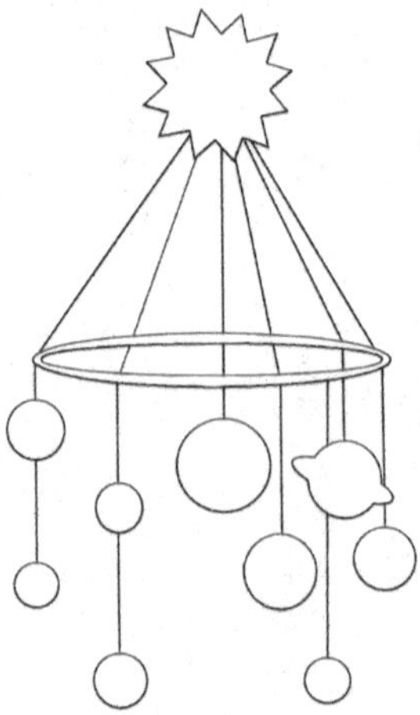

Also, by Christopher Cotton:

We were always unravelling.

Frayed is a quiet unpicking of what keeps us together – a thematic collection of poems and prose that traces our collective journey through evolution, erosion, and awakening.

Reflecting on complacency, digital estrangement, and the stories we tell ourselves to cope, this work offers not solutions, but clarity: a call to notice what we've accepted, what we've lost, and what we might still reclaim.

What happens when we truly listen to the other side, of an argument, of an identity, of ourselves?

Encounters is a poetic exploration of contrast and connection, pairing twenty-five perspectives across lines of belief, experience, and form. From the riotous streets to quiet bedrooms, from machines and moonlight to parents and teens, each poem offers a mirrored voice, not in competition, but in conversation.

We are not the sum of our circumstances. But we are shaped by them.

Situations is the final part of a quietly powerful trilogy exploring what it means to be human in a world that misjudges, misnames, or misunderstands us.

Caricatures is a poetic parade of stereotypes - from cradle to comment thread.

With biting wit and unexpected warmth, Christopher Cotton casts his eye over the familiar figures of modern life: the school-run mum, the remote worker, the motivational poster believer, the troll. Each verse is a snapshot of a world we all recognise... sometimes too well.

What is a pest? What makes something vermin? And who gets to decide?

In this powerful and poetic journey through time, Christopher Cotton traces the evolutionary arc of the animals we most often seek to trap, banish, or exterminate. From the flea to the fox, from the ant to the ape, *Vermin* explores what it means to survive, and to be judged for doing so.

Christopher Cotton

ISBN: 978-1-9192021-1-2

www.ingramcontent.com/pod-product-compliance
Lightning Source LLC
LaVergne TN
LVHW040520210926
848480LV00007BA/1525